COUNTRYI
HERITA

COUNTRYMAN'S

HERITAGE

By
David Obbard.
Sussex Farmer
and Pastor.

"The lines are fallen to me in pleasant places;
yea, I have a goodly heritage." Psalm 16 v6.

Articles, Anecdotes, Pictures and Poems,
with country wisdom, which is part of the
countryman's goodly heritage.

Cherry Gardens Publications,
Groombridge, Tunbridge Wells.

Acknowledgements:

R. Obbard; pictures	No.	14, 24, 30a, 48, 43a, 54, 80.
J. Pratt;	No.	8, 70.
D. Honeysett;	No.	56b.

Sue Obbard for pen and ink drawings.

CONTENTS

JANUARY

THE COTTAGE LOAF and THE HELPING HAND

IN the old cottage where my father and grandfather were born is the remains of an old brick oven. It is simply a large cavity, lined with bricks, in the wall at the side of an open fire place called an 'ingle-nook'. On baking days my grandmother would light a fire inside her oven, using only dry faggot wood (that is, bundles of small sticks) and when the bricks got hot would rake the ashes to the front edge, sweeping the floor clean with a few twigs of green broom. She would then get the food she had prepared in 'the baking table corner' and place it in the oven, putting an iron door up to the front. By holding her hand in the oven she would have judged the heat, and known how long to leave the food before taking it out with a long handled bread shovel.

My mother's mother had no bread oven in her cottage at Motts Mill, and the baking was done by her husband's mother who lived at 'Broomfields', not far from Cherry Gardens. This meant that to collect the bread someone had to walk about a mile along a narrow road and across the fields. One night my mother and her brother had been on such a journey when the snow lay thick on the ground. As they walked up the hill near their home there was a noise in the hedge, and a great flurry of snow came down the bank. They turned tail and fled, being very frightenend.

At the bottom of the hill they ran into the doctor, who had just been to see a patient and was walking back to Groombridge. "Hello my dears", he said, "What's all the hurry?" Breathlessly they explained, "We think there's a man in the hedge ready to jump out at us". "I don't think so", said the doctor kindly, "you come along with me and see." So, with their hands in his, they bravely marched back up the hill, and got home safe at last.

Winter Sunshine. page 9.
Pond and oast houses at Hendal Farm, Groombridge.
(Hendal Manor is mentioned in Doomsday Book)

Perhaps the majority of those who read this will never have to walk in dark lanes at night. If any do, it is doubtful whether they would find a doctor to hold their hand. But God has given us a wonderful promise; "I the LORD thy God will hold thy right hand, saying unto thee, Fear not; I will help thee" (Isaiah 41: 13). Young and old, we may go safely and bravely through life knowing God will fulfil this promise for all those who put their trust in him. Trusting in that gracious and almighty hand we shall arrive safe Home at last.

'If I could find some cave unknown,
Where human foot had never trod,
Yet there I should not be alone
For all around me would be God.'

WINTER SUNSHINE

HOW good it is to see and feel the sun. Even on a winter's day it comes as the promise of summer. It will not be long before we see the buds begin to move, the flowers appear, and a multitude of signs that spring is on the way. God's promise to Noah, of seed time and harvest, summer and winter, cold and heat, is constantly being fulfilled, and we thank him for this.

We too make promises, but are sometimes unable to fulfil them; perhaps we forget, or changing circumstances make it that despite our good intentions, we miserably fail. God gives us assurance that what he has promised, he is able and willing to perform. When my son was about twelve years old, somebody asked him, "Can your Dad sharpen saws?" He replied, "My Dad can do everything!" Alas, as he grew older he found that absolute confidence in his father was not justified, but the Christian has a Father who can indeed do everything. Many there are who have found Jesus' words true, not only in regard to salvation, but in many practical matters; "With men it is impossible, but not with God: for with God all things are possible".

When we remember that this God has promised to hear and answer prayer, and that he will listen to a child, this encourages us to ask him to help us in matters small and great. King David said, "I love the Lord because he has heard my voice and my supplications; therefore will I call upon him as long as I live". That was first a good testimony, and then a good resolve. If we make and keep this resolve, then King David's testimony will also become our own.

God's promises are ever sure,
No circumstance can thwart his will;
His love and purpose will endure,
Then let your troubled heart be still.

CREATION and RE-CREATION

PSLAM 19 reminds us the glory of the Lord in creation is something everyone who has eyes can see, - 'The heavens declare the glory of the Lord.'

That glory means different things to different people; for example, the artist, the meteorologist, the astronomer and the astronaut all view this with different thoughts than David expresses in Psalm 8: v3. As a Christian farmer and horticulturist, I see and appreciate this glory perhaps more than many, not only in the heavens, but even in the earth under my feet.

Although not qualified to tell how many million living creatures and organisms there are in a cubic metre of soil, I see something of the glory of God in that he has given 'herb for the service of man, that he may bring forth food out of the earth' (Psalm 104; 14). The farmer today does not have the same contact with the earth that I had as a young man walking barefoot in the freshly turned furrow behind a horse-drawn plough. The touch, smell, aesthetical quality and joy that is associated with such a life has been experienced by many, even atheists, but for me the glory of God is there.

That this soil, sometimes frozen too hard for a pick-axe to penetrate, or too wet and soggy to bear the weight of a dog, should, under the influence of the seasons as described in Psalm 65 turn to the soft crumbly texture required for sowing the seed in spring-time, never ceases to amaze me. No effort of man, even with modern machinery, can produce the same results.

In the same way God is able to take the most unlikely and unlovely characters, or the most hardened of criminals, and under the influence of his Holy Spirit completely change the texture of their lives. The hard heart becomes a well-watered garden, and the wasted life a fruitful field.

THE ANOINTED HEAD

A FARMING friend told me that when his father came up from Devon to farm in Wiltshire, he bought two rams to run with a flock of ewes on the golf course and Down lands which were on the edge of his farm. He was not a 'sheep man', and soon learned an expensive lesson, for the two rams fought for the place of 'top-sheep', and in the process one was killed. Although appearing timid and harmless creatures, sheep, especially rams, fight by charging head-on at each other, when their thick skulls crash together with tremendous impact. These two rams, whilst fighting, had collided with such force that one had a broken neck.

On learning of this a neighbour said, "If I had known you were putting those two rams together I would have told you to rub their heads with lard." Liberally greasing the head and nose of each ram would have the effect of causing them to slide off each other in a ridiculous way, making them feel frustrated, and thus releasing the tension.

The Anointed Head. page 11.
Sheep grazing beneath a Sussex windmill, with the shepherds hut in the background.

The Bible describes us as sheep, and this is one of the lessons we need to learn from them. How often we clash with others, sometimes through trying to assert ourselves and become 'top-sheep'. Much hurt, injury or bruising, if not worse, may result from this. How we need the Good Shepherd to anoint us with his Holy Spirit, and the grace of humility, to keep us from acting like these silly sheep. I wonder if this was one of the things David the psalmist was thinking of when he wrote, 'Thou anointest my head with oil'. The writer of Proverbs evidently had the same principle in mind when he said, 'A soft answer turneth away wrath.'

FEBRUARY

DARKNESS AND LIGHT

WHEN I was a little boy I was terribly afraid of the dark. There were no electric or gas lights in our home, and as my father was not very sympathetic about such things he would say, "If you see a bogey, bring him to me; I will paint him white, and then we shall all be able to see him." This did not help allay my fears when going upstairs to bed, or out to the very primitive toilet in the garden, without a light on a dark winter's night. As I grew older I realised there was nothing to be afraid of, but here is a lovely story of how one little girl overcame the same fear.

The story is of a mother and her four year old daughter who were preparing to go to bed for the night. The child was afraid of the dark, and being on their own on this occasion, the mother was also apprehensive. When all the lights had been turned out the little girl noticed the moon shining brightly through the window.

"Mummy", she asked, "is that God's light up there?" "Yes it is dear," came the reply.

Darkness and Light. page 13
An electric storm over the town of Hastings.

"Will he put it out and go to sleep too?" was the next question, and mummy said, "Oh no: God never goes to sleep."

For a few moments there was silence, and then the little one said thoughtfully, "Well Mummy, as long as God is awake, I am not scared." Realising that the Lord would be watching over her, the reassured youngster soon fell into a peaceful sleep.

Christians may confidently commit both day and night to their ever faithful God. He is fully aware of our fears in the dark, as well as our frustrations and needs in the light. We can be assured of his constant care, for his loving eye and protecting hand are always on us.

"He who keeps you will not slumber.
Behold, he who keeps Israel shall never slumber nor sleep."

PULLING TOGETHER

MY great-grandfather was a carter on a neighbouring farm. Like all those of his day, his life centred round his team of horses, and there was a remarkable bond of understanding between man and beast. The men were proud of their horses, and there was often friendly rivalry between the local carters over their team's appearance and achievements. My father told me the following story:

It was the custom for farmers in this area to go to Lewes for lime to put on their fields. As Lewes was nearly twenty miles away, this entailed an early start for the men, and a long hard haul for the horses pulling the large wagons. On one occasion my great-grandfather took a team of three, one horse in the shafts, and two trace horses out in front. At the chalk pits a neighbour's team of four were unable to move their loaded wagon which had become bogged down in a gateway, and the carter asked for the two trace horses to be added to his own.

The Never-ending Miracle. page 7.
House and kitchen garden at Cherry Gardens. The barn in the rear has a flail threshing floor.

"No!" said the old man, "take your four off, and I will pull it out with my three." After harnessing his own horses to the wagon, he stood well back from his team, where each horse could see him, and began slowly to raise his whip. The first horse took the strain by leaning into his collar until the tug chains were tight. As higher went the whip, the second, and then the third horse, did the same. A word of command, and a crack of the whip, and all three horses pulled as one, bringing the heavily laden wagon out of the hole and up onto the road.

Those horses were no bigger or stronger than the others, but they had been taught to pull together. Surely there is a lesson in this for us in our families and churches. How often little is accomplished because members do not pull together. The psalmist of old knew the secret; "Where the brethren dwell together in unity... there the Lord commandeth the blessing".

THE NEVER-ENDING MIRACLE

"HE causeth grass to grow for the cattle, and herb for the service of man (plants for man to cultivate), that he may bring forth food out of the earth." Psalm 104.

Food out of the earth: the soil. Never refer to soil as dirt! Only man makes dirt; he alone of all God's creatures pollutes the environment. The earth tells of a God never wanting, never wasting, as all excess is recycled. Next time you pass a good fruiterer's or green-grocer's shop, just pause and consider - it all comes out of the earth. Fruit and vegetables from all over the world are seen in their abundance, various flavours, colours, smells and nutritious values. They are displayed to 'catch the eye' in order to 'empty the pocket'. To me they display the wisdom, power and glory of God, who has enabled some man somewhere to take part in the never-ending miracle of bringing forth take part in the never-ending miracle of bringing forth food out of the earth. Even if man could invent a machine which could turn soil and water into food, it would not compare with God's creation plan, for the soil would be used up in the process. Only a Creator could bring it out of the earth.

I live in a little cottage where my grandfather was born. He, like his father and grandfather before him, brought forth food out of the earth in the garden. My father used the same soil, and I do so still today. Unless some man kills the soil with man-made chemicals it will continue to be productive until the Lord returns. It is of this never-ending miracle that the psalmist speaks in Psalm 104; 30-31.

"Thou sendest forth thy spirit; they are created;
And thou renewest the face of the earth.
The glory of the Lord shall endure for ever;
The Lord shall rejoice in his works."

So long as God spares my life on this earth, I too shall rejoice in them and give to him the glory.

SHEEP OWNERS AND SHEPHERDS

SPRINGTIME brings the lambs. Over the recent years one of the changes in agriculture brought about by the financial situation in the European Common Market has been a marked increase in the number of sheep in this country. At one time confined mainly to certain areas in the British Isles, such as the hills, downs, and marshes, they now feed in many places where once only cattle grazed.

In Isaiah 40 we read of God, "He shall feed his flock like a shepherd; He shall gather the lambs in his arm, and carry them in his bosom." The loving care of an eastern shepherd is proverbial, and still today the dedication of a true shepherd is seldom matched in any other sphere of country life, but, sad to say, not all who own sheep are true shepherds.

I have never had sheep of my own, but have from time to time taken in 'keep sheep', that is, sheep which have been grazing on the marshes during the summer, and brought on to dry, well- drained land during the winter months. Once I phoned the owner about a poorly lamb which was in our field; all he said was, "Ah well, we expect to lose a few." He showed no real concern because he considered it would cost more to save its life than the lamb was worth.

What a contrast to this farmer we have in our God. Jesus said, "I am the good shepherd; the good shepherd giveth his life for the sheep." Although we were sinners, He died to save us from our sins. Despite all care and precautions even the best shepherd must expect to lose a few; only Jesus could say, "I give unto my sheep eternal life; they shall never perish, neither shall any man pluck them out of my hand."

"THE SPIDER takes hold with her hands, and is in the king's palaces."

What perseverance spiders show
To get their webs made fast:
No matter how the wind may blow,
They get them fixed at last.

And though it is a tiny thing
(Which some cannot endure),
Yet in the palace of the king,
It finds a place secure.

Then let us also persevere,
That God our souls may bring
To enter, free from sin and fear,
The palace of THE KING.

Big Jugs and Little Jugs. page 21.
At the door of Danemore Farmhouse, Speldhurst.

MARCH

LITTLE JUGS AND BIG JUGS

WHEN my mother was a little girl she was sent to the neighbouring farm to get the milk - one pint for a penny. (240 pints for a £1!) Being a sharp-eyed little character she soon noted that the boy from next door got a large jug full for a penny, so up she spoke. "Why does he get more for a penny than we do?" The farmer's wife smiled and said, "Because he brings a bigger jug."

There is a lesson for us there. In the Epistle of James we read, concerning prayer, "Ye have not because ye ask not". Very often the people who can testify to wonderful answers to prayer are those who act upon God's promise, "Open thy mouth wide, and I will fill it". They take a bigger jug! "Thou art coming to a King, large petitions with thee bring, For his grace and power are such, none can ever ask too much."

There is an equally important lesson to be learned from the sequel to this story, for the next day my mother said to her elder sister who acted as mother to the family when their own mother died, "Annie, I'm not taking the pint jug today, I want a bigger one."

"What do you want a bigger one for?" inquired her sister. "Dennis gets a big jug full for a penny, so I'm taking our biggest jug too."

"No! that you don't my girl", said Annie, "We only pay for a pint, so we will only take a pint."

Much as we commend the 'poor but honest' attitude displayed by this answer, might there not have been a little of the 'poor but proud' in it also?

God has exhorted the thirsty to "Come, buy wine and milk, without money, and without price". He has said, "Whosoever will, let him take of the water of life freely". Sadly many are too proud to own their spiritual bankruptcy and seek the blessing which comes to us so freely through him who "was wounded for our transgressions and bruised for our iniquities". What a large jugful, an overflowing cup, many have received when coming in the spirit expressed by Toplady's hymn:

"Nothing in my hand I bring,
Simply to Thy cross I cling;
Naked, come to Thee for dress,
Helpless, look to Thee for grace,
Foul I to the fountain fly,
Wash me Saviour, or I die."

UNSEEN PROTECTION

"THEY that trust in the Lord shall be as Mount Zion, which cannot be removed, but abideth for ever. As the mountains are round about Jerusalem, so the Lord is round about his people, from henceforth, even for ever." Psalm 125. 2.

Some years ago I saw a perfect illustration of the Lord's unseen protection of his people. It was of a cranefly, or 'Daddy-long-legs', flying frantically up and down inside one of our windows in a great fright, whilst outside was a sparrow, likewise flying up and down, pecking hard as it tried to reach the fly inside.

The cranefly did not see the glass, but it saw the sparrow, and evidently expected every moment to be caught. Neither did the sparrow see the glass, but it saw the cranefly, and was sure it would catch it. Yet all the while, because of that thin, invisible sheet of glass, the fly was perfectly safe.

How often we, who have trusted the Lord as our Saviour, forget he is also our Protector and Provider, and our hearts fail us. We read of how Elisha's servant was in great fear when he awoke in the morning and saw the city surrounded with horses and chariots, all the host of the king of Assyria. But when his eyes were opened at the prayer of the prophet, his fears vanished, for he saw; 'and behold, the mountain was full of horses and chariots of fire round about Elisha'.

If we remember that "The angel of the Lord encampeth round about them that fear him, and he delivereth them", This will deliver us from frantic fear, and crippling anxiety.

"A sovereign Protector I have,
Unseen, yet for ever at hand;
Unchangeably faithful to save,
Almighty to rule and command."

A LESSON FROM A BIRD.

WHEN spring is on the way, with flowers in the gardens, and trees beginning to show their leaves, we may expect to find the birds making their nests. One of the first signs is to see the rooks congregating in the tops of the trees where they have their rookery, making a great fuss over renovating what the storms have left of last year's nests. Not many birds in this country form a little colony, or group of nests like the rooks. Most prefer to hide away in some sheltered spot where they can rear their little ones in peace.

Every year we find one or two nests in the fruit bushes, or tops of the raspberry rows, sometimes so well hidden they are not found until the leaves fall in the autumn. When discovered by the people who come to pick the raspberries, these nests are a source of great interest and pleasure. It is surprising how trustful a wild bird may sometimes be. Some years ago we had a greenfinch which nested in the same gooseberry bush for several seasons, and many times I fed that little bird with small caterpillars.

One of the most remarkable stories I have heard concerning a bird on a nest was told me by a friend who saw action with the Eighth Army, fighting in North

Saving the Cat. page 25.
"Curiosity killed the cat." A group of kittens display their natural inquisitivness.

Africa. On one occasion his tank received a direct hit, killing the driver, and forcing the crew to leap out and seek cover. As he did so my friend saw, in a bush, surrounded by the din of modern warfare and burning tanks, a bird sitting on a nest. That little bird reminded him of the perfect peace which God has promised to those who stay their minds upon him. There, in that battlefield, God enabled him to put his trust in the Lord.

We hope that we will be preserved from war, so that our young men will never find themselves in the same situation as my friend, but I pray that all who read this story will come to know 'the peace of God which passeth understanding' by trusting in the Lord Jesus Christ as their Saviour.

SAVING THE CAT

"CURIOSITY killed the cat" is an old saying, the truth of which I have often seen demonstrated. On one occasion, some fifty years ago, my father and I heard my Aunt Ada, who lived next door, calling for help. We rushed from our work and found her, with uncle George and his workman, peering down into the well at a young cat, whose curiosity and playfulness had caused it to make one leap too many. There it was, some 25 - 30 ft. below us, frantically struggling in the water.

After a hurried consultation my uncle, who was a very short man, volunteered to go down, placing one foot in the bucket and holding on to the chain. Very carefully he was lowered until he could rescue the cat from its watery grave, and then with steady turns of the handle Ted began to rewind the chain on the roller to bring up its precious burden. We watched with bated breath for uncle to emerge from the well, and heaved a great sigh of relief as he stepped safely onto solid ground, with the cat held securely in one arm.

Probably the incident would have been forgotton if it had not been for the fact that as I was drawing water a few days after this had taken place, a link in the chain snapped, and the bucket went crashing down to the bottom. "Gosh!" I thought, "If that had happened the other day, uncle George would have been killed: he risked his life to save a cat!"

Many times since I have thought of this incident as a beautiful illustration of what the Lord Jesus did. He came down to earth to save us when we, like the cat, were in a position where there was nothing we could do to save ourselves. He not only risked his life, he actually died to save us, and this is why the resurrection is such an important truth.

If my uncle had himself perished in the well, there would have been no salvation for the cat. When he reappeared we knew that the cat was safe also. Our Saviour's resurrection is a proof that his mission of salvation was successfully accomplished. It is also important to notice that the cat had to trust my uncle. It is true that Jesus "died for our offences, and rose again for our justification", but the Bible also teaches that we cannot be saved unless we believe and trust in him.

"Lo, Jesus meets us, risen from the tomb!
Lovingly he greets us, scatters fear and gloom;
Let the church with gladness hymns of triumph sing,
For her Lord now liveth, death has lost its sting.
  Thine be the glory, risen, conquering Son,
  Endless is the victory Thou o'er death hast won.

No more we doubt Thee, glorious Prince of Life;
Life is nought without Thee: aid us in our strife;
Make us more than conquerors, through Thy deathles
  love
Bring us safe through Jordon to Thy home above.
  Thine be the glory, risen, conquering Son,
  Endless is the victory Thou o'er death hast won."

APRIL

A GLIMPSE OF GLORY

AS one who sees much of God to admire in nature, and many reasons in the natural realm to call for his praise, I cannot understand the view inferred by some that such worship is not "spiritual". Nothing could be more spiritual than the inspired Psalms, many of which major upon the worship of God as Creator. Romans 1: 19-21 clearly indicates that such worship is always acceptable to him.

There are many ways in which the worship of God, a sense of his greatness and glory, may be inspired through nature. There can surely be few Christians who have not at times been affected by the beauty of the sunset, the delicate pattern of a wayside flower, the majesty of a star-lit sky, or the awe- inspiring display of an electric storm. In such moments we worship God as Creator.

Sometimes it is the unusual or unexpected which catches the eye; at other times the most common and humble thing may move the soul to adoration. I shall never forget one morning as I returned from milking the cows, pausing in a field of barley. I kneeled down to inspect the tender green shoots more closely, and as I looked up the whole field was transformed before my eyes. A tiny spider's web appeared to float from the tip of every blade, reflecting the light of the rising sun, so that the area became a shimmering sea of silver.

There came upon me such a sense of the glory and beauty of God in his creation as I could not find words, or even thoughts to express. My whole being was deeply moved as I sensed the very presence of God, and knelt in humble awe before my Maker.

ENTANGLED

FARM cats! What a variety we've had, and how far and wide have their offspring migrated, especially through some child being enthralled by a kitten. Baby kittens are irresistible and the mother cat knows this, often hiding them away from prying eyes. She seems to know that although little children love kittens, their clutching hands would hurt her babies, and their voices would frighten them. When the kittens are bigger and come out to play they are a constant source of pleasure, even if they do get up to mischief. My wife's mother used to sit and knit, and sometimes a playful kitten would pounce on her ball of wool. The kitten usually had more fun than Grandma!

One day a kitten found a ball of wool on a chair. It was just inquisitive at first, a little unsure what it was, and gave it a pat. A strand of wool caught in its claw, and as the kitten shook its paw the ball moved. Growing more bold it patted again with its paw, and the ball rolled onto the floor. Soon it was having great fun, chasing the ball, lying on its back kicking, and making a terrible mess of tangled wool everywhere. Of course the wool not only wound round the chair-legs, but round the kitten as well, until a strand of wool pulled tight round kittie's throat. What began as a game of fun became a fight for life, but the more she tried to escape, the more tightly she became bound. At last the kitten was so exhausted that it lay and cried feebly. The mother cat was unable to help, but someone else heard its cry, and coming with a pair of scissors soon set it free.

There is a lesson for us all in this. Sin, which means to come short of the standard of life God requires of us, is like that ball of wool. All unthinkingly we may become entangled. What at first seems quite innocent, even "good fun", becomes a snare from which neither we, nor any other person, can set us free. Thanks be to God, when we realise our struggles are useless and cry to Him for help, there is One who can. The Lord Jesus Christ said, "If the Son shall make you free, you will be free indeed". Let us cry to him as soon as we are aware of the 'smallest' wrong thought or action, and we shall know what it is to be "Kept by the power of God."

"Let worldly minds the world pursue,
It has no charms for me,
Once I admired its pleasures too,
But Jesus set me free."

THE UNFAILING GUIDE

THE study of migrating birds and other creatures is a strange and fascinating one. How wonderful that God has given to them this instinct, by which they are able to travel distances of many thousands of miles, arriving in the right place at the right time. Year after year swallows return to the same nesting place in this country, having spent the winter months around the Mediterranean or in South Africa. We are always pleased to see them, and hear them twittering away as if to say, "here we are, we're back again." My little grand-daughter called them "Tweety-zoomers" as they swept down to frighten the farm cats.

Another spring visitor is the cuckoo, which we usually hear in the middle of April. Early in the morning he calls "cuckoo" to all we foolish human beings who tend to lie in bed when the sun shines, after sitting up at night using electric light! In some ways the journey of the cuckoo is more wonderful than that of the swallow, because all the adult birds leave this country before the young ones, who then have to find their way without the more experienced ones to guide.

God made provision for all these birds and creatures to go to their 'better country' by giving them instinct, which means 'urged from within'. To guide us to our

The Unfailing Guide. page 29.
Swallows on the nest, ready to fly.

Pulling Together page 15.
Working with a springtine harrow.

"better country, that is an heavenly"(Hebrews 11;16), he has given us his holy word and his Holy Spirit to teach us the way, having also given his Son to be our Saviour. However, let it be noted God does not carry these birds, though they often have a wind to assist them - they have to fly.

"Flee to him, your only Saviour,
In his precious name confide;
In the whole of your behaviour
Own him as your sovereign guide."

AN IMPORTANT VISITOR

ONCE again in God's wonderful plan of renewal, "The flowers appear on the earth, and the time for the singing of birds is come." The flowers that brighten our gardens and give pleasure to our eyes are not only things of great beauty, but of great potential. What happens to the flower has a marked effect on the next generation, or the future of the species. If it is robbed by an earwig, and unvisited by the bee, it may simply brighten the earth for a little while, and die without having produced any fruit or seed. There are some flowers which have the capacity to shut up the way to their vital centre when it is dark, and only open to the sunshine when the bees are able to fly.

God's word describes man as the flower of the field. What a blessing when God's great fertilising agent, the Holy Spirit, visits this flower, especially when it is young, and imparts spiritual life. We pray for the young people of our day that they may guard the sweet nectar of youth from all evil things by closing their eyes, ears, and minds to them, having their hearts opened to the word of God. Many great men and women, whose lives have been fruitful in various fields of service, were what they were because in the flower of their youth they were visited in this way.

Blackthorn Winters. page 33.
Blackthorn and daffodils in full bloom.

MAY

BLACKTHORN WINTERS

WHEN we get to the first of May we can say with certainty that spring is here. Everywhere we look in the countryside there are signs of life transforming the scene, and showing the wonderful handiwork of God. All along the hedgerows, and in the woods, can be seen a hundred shades of green, for even trees of the same sort vary, yet all are in harmony with oneanother.

Many of the flowering shrubs are already in bloom, and in a short while the apple blossom will burst in a riot of colour. All over the countryside where I live appear patches of white, shining in the hedgerows - the blackthorn in full flower. This bursting into flower usually comes at the same time as a sharp spell of cold, easterly winds, frost, or even snow. To country folk this time is known as 'Blackthorn Winter'.

Blackthorn winters appear in our lives also. As the warm days of April bring on the young growth and flowers, so we often see signs of spiritual life, perhaps in children, and hope to see the bud of grace come to perfection, but some chill wind or sharp frost comes and 'sets things back'.

Sometimes those who have been seeking the Lord, and feeling the growing awareness of spiritual life, find that, instead of blessing, they come into the experience Newton describes in the hymn which commences: "I asked the Lord that I might grow in love, and faith, and every grace". The Lord answered that prayer by making him feel 'the hidden evils of (his) heart', and not, as Newton hoped, in some dramatic 'favoured hour'.
Blackthorn winters, whether in nature, our providential paths, or in the things of the spirit, have their place, and are followed by sunshine and flowers. May God bless all the 'set backs' that come into our lives to make us more fruitful.

The Cottage Loaf. page 7.
A corner of the inglenook, showing the position of the bread oven. The straight joint in the brickwork reveals this to have been built in at a later date.

The Bread Oven; later alterations to the house cut through the oven itself. Note the brickwork blackened by smoke which went up the open chimney.

THINGS NEW AND OLD

THERE are two things which always impress me at springtime. One is the newness, the wonder of what we see around us. The leaves upon the trees unfolding in their many shades of green, the riot of colour in the flower beds, the garden shrubs, and the golden glory of the gorse bushes on Ashdown forest. We have seen it all before, yet it is new and fresh and wonderful, stirring something in our hearts as though we saw it for the first time.

The other thing is that although it is the same season, and the same scene, it is always different. Sometimes early, sometimes late, sometimes prolonged, with the early flowers over before the later ones appear, and at other times the peonies coming into bloom while the daffodils are still with us. Those who search the hedgerows and country lanes make fresh discoveries of violets, orchids, and other plants in unexpected places.

To those who know and love the Lord Jesus, it is the same with the things of God and the preaching of the gospel. We have 'heard it all before', and perhaps at one time wondered why people went week after week to listen to the same man preaching about the same things. Now we find something ever new in the 'Old old story of Jesus and his love.' We want no new gospel, only the old gospel made new.

At the same time, however familiar we are with a passage of scripture, there are always fresh things to discover; aspects of truth we have not seen there before. These things gladden our hearts and refresh our spirits as we pass along the road of life.

"The kingdom of heaven is like a man that is an householder, who brings forth out of his treasures things new and old." (Matthew 13;52)

KNOWING THE UN-KNOWABLE

MAY time is flower time! It is not unusual for ornamental flowering shrubs to bloom in the 'wrong' season if the weather is warm, but fruit trees always wait until the right time. That 'right time' is mysteriously linked with Easter, whether it comes early or late.

There are also many wonderful things in nature we cannot understand. I have often pondered over this little poem.

"I cannot understand the reason why
The seed I plant, so very brown and dry,
Warmed by the earth, and fed by sun and dew,
Can grow to perfect flowers of every hue.
I cannot understand - but this I know:
God makes them grow.

Or how this speckled egg, so sweet and small,
That lies within my hand, a shining ball,
Can hold a living bird, to chirp and sing,
And sweep from tree to tree on graceful wing.
I cannot understand - but this I know:
God made it so.

But higher still than these, I cannot see
How such a wondrous thing can ever be...
That if I come in faith, and humbly pray
The sinner's Friend to take my sins away,
The Lord of heaven, of flowers and birds will come
And make my humble heart his daily home.
I cannot understand - but this I know:
It will be so."

WE would think it a great honour if the Lord Jesus were to come and stay in our home, but he has promised to dwell in our hearts. By his Spirit we may know his love,

Knowing the Unknowable. page37.
A border of annual flowers at Drusillas, Sussex.

Things New and Old. page 36.
Flowering shrubs and conifers in the rockery at Leonards Lee.

help and guidance every day, just as surely as his disciples did when he was here on earth. We pray especially for young people, that they may know him in the springtime of youth, so that before the autumn of life comes they may be mature and fruitful Christians.

GREEN FINGERS

SOME people have a wonderful aptitude for growing things - what the old folk call "having green fingers". My mother was such a person. Her garden was always a riot of colour, with something of everything in what appeared to us utter confusion, although she knew where every precious plant was. Many of them came from small pieces which she had literally 'pinched ' off her friends bushes (with their permission), and simply 'stuck them in' where they grew and thrived.

She also had a great collection of indoor plants, including some flourishing maidenhair ferns. One of these she gave to my sister, who was at that time nursing in Gloucester. Taking my mother to see her on one occasion I heard and witnessed something which reminded me of Isaiah's words concerning Messiah, the Lord Jesus Christ. "He shall grow up before him as a tender plant, and, as a root out of a dry ground, he hath no form nor comeliness; when we shall see him, there is no beauty that we should desire him." (Isaiah 53; 2)

"Where is that maidenhair fern I gave you Ruth?" - "Oh, it died."

" But where is it?" - "I don't know: I threw it out somewhere."

A search by my mother eventually located it, thrown into a corner, looking as dry and dead as anything could be. After emptying it out of the pot and shaking off all the soil from the roots, she examined it carefully, and said, "I think there might be a little life left in it." We brought it home, and soon under her expert care there appeared a 'tender shoot', and in due time the plant was restored to its former glory.

Alas, to how many does our Saviour still appear to be 'a root out of a dry ground'. Christ is cast out as my sister cast out the maiden-hair fern, without any thought, understanding, or real examination. Yet, just as a root out of a dry ground may have life and potential unthought of by many, so the Lord Jesus has 'life in himself' which he imparts to others, that they too may be restored to glory. He then is no longer 'without form or comeliness', but 'the rose of Sharon, the Lily of the Valley, yea, he is the Altogether Lovely'. Which is he to you?

Thoughts prompted by seeing a cow following her calf.

The cow was big, headstrong and wild;
No man could ever hold it:
But see, it follows like a child!
A wonder to behold it.

What has its nature wild subdued
That it will roam no longer?
Her gaze upon her calf is glued;
Maternal love is stronger.

She's bound by cords we cannot see,
They govern her direction:
So God's great love and wise decree
Has won our hearts' affection.

The Saviour seeks us from above,
He draws, but drives us never;
Loved by his everlasting love
We're bound to him for ever.

Reflections. page 83.
Old Scotney Castle, near Lamberhurst, Sussex.

Good Figs and Naughty Figs. page 44.
Strawberries, just right for eating with cream.

Losing the Load. page 69.
A corner of the Sunday School Harvest Table.

GOOD FIGS and NAUGHTY FIGS

JUNE to July is strawberry time! One of our customers who comes daily for strawberries throughout the season said, "I don't freeze them, or make jam of them, or anything silly like that. I just eat them with sugar and cream - that's my idea of summer." She is not alone in thinking there is nothing better in the way of fruit than a lovely strawberry straight from the plant and dipped in sugar. However, when a strawberry has 'gone off', as they may quickly do in hot weather, there is nothing more unpleasant. There are two lessons we may learn from this.

First, the better a thing is when it is good, the worse it becomes when bad. The Lord showed Jeremiah two baskets of figs. "One basket had very good figs, like figs that are first ripe; and the other basket had very naughty figs, which could not be eaten, they were so bad." Jer. 24;2. The more talents and gifts a person has, the more good they can do if rightly used, and the more evil they become if that one goes away from God.

Secondly I notice that if a strawberry has only the tiniest spot of rot in it, the whole strawberry tastes horrible. This is because the spot on the outside is an indication that it has gone wrong within. That reminds us that our sins, no matter how small they may appear to us, are an evidence that our heart is not right before God. The Bible tells us that "The heart is deceitful above all things, and desperately wicked."

There is nothing we can do to make a bad strawberry into a good one, but God has power to transform the worst of characters into true Christians, "having favour of God and man". He has promised to give his Holy Spirit to all who sincerely ask, and he can enable us to use our talents, whether few or many, to his glory and the good of our fellow men.

JUNE

THE SHELTERING WING

WE no longer raise chicks, or keep hens on a commercial basis, but we do have a few hens of different varieties running free in the 'Pets Corner' on the farm. Not infrequently one will disappear for a time during the early summer, only to turn up three weeks later with a brood of little chicks trailing along behind her.

The lovely picture of chicks being protected by a mother hen was used by Jesus as an illustration. When danger threatens, the mother hen 'clucks' or calls her little chicks. They will immediately obey the call, and run to her for shelter and safety. Years ago we used to raise hundreds of chicks in this way, and I often saw this happen as a hawk hovered over the field in which they were. Any chick which refused to obey her call would be in danger, and sometimes the cruel hawk would swoop down and kill one before its mother's eyes.

Jesus wept over Jerusalem, even while the people were rejoicing and singing "Hosanna to the Son of David", because he knew what was going to happen to them. Earlier he had said, "How often would I have gathered thy children together as a hen gathers her brood under her wings, but ye would not." Although the little chicks run to the hen when she calls, many who heard the call of Jesus despised him, and his gospel. They "would not come unto him that they might have eternal life." John 5;40

What a blessing awaits those who obey the gospel call, and flee from the dangers that surround them to God as David, who said, "Under thy wings will I rejoice." Satan, like a great hawk, is ever watchful, seeking to destroy the souls of men.

"He that dwells in the secret place of the Most High,
Shall abide under the shadow of the Almighty."

The Wick Lamp. page 77.
This stands in what was the inglenook window. Note the wooden yoke in the gun rack, and the wooden-framed bow saw in the old seat recess.

The Aladdin Lamp page 78.
A recently built brick flue now takes the smoke from the fire which burns on the hearth. The brass bed-warmer on the left was filled with hot ashes.

The Important Visitor. page 31.
A honey bee visits a messembryanthemum flower.

THE BOTTLES OF HEAVEN

IT was one of those 'catchy' summers, when haymaking was difficult. So often just as the grass was ready to bale, down came a shower, and the whole operation had to stop. We had a small field down and had been constantly turning it in the hope of snatching it from the weather, and as the day dawned bright we made a start. Alas, only too soon the clouds gathered, and we worked hard, one man baling, the rest picking up and carting to the stack.

We could see the rain sweeping towards us, and I was asked, "Hadn't we better pack up before we get soaked"? Needless to say, I was praying. Prayer, as Jesus said in John 4, is not confined to time and place, and I found the simplest operation became a matter for prayer; how much more then, faced with such a situation. In the next field a neighbour was also busy haymaking, and as the rain came down he had to leave off work and run for shelter. Miraculously the rain did not reach us.

Throughout the day the showers of rain continued to fall, sweeping through the valley below, or turning away from the hilltop on which we were working. In mid-afternoon we collected the last bales and headed for home. Just as we had finished unloading, and got out the waterproof stack sheet, down came the rain! As I tied the ropes to secure the sheet, I was soaked to the skin. Once again he who 'holds the waters in the hollow of his hand' had 'stayed the bottles of heaven' in answer to prayer.

Although the short term weather forecasts have become much more reliable since the use of computers, it is a mystery to some why, with weather reports coming in from space and every spot on earth, long term forecasts are still highly inaccurate. Might it not be that it is because the meteorologists cannot put into the computer the most important factor of all - that he

Cow and Calf. page 41.
No machine milking here!

who formed the world, and laid down the laws which govern all that is within it, including the weather, is still in control? This is so beautifully described by the psalmist in Psalm 107, concerning those that go down to the sea in ships.

"Then they cry unto the Lord in their trouble, and he bringeth them out of their distresses. He maketh the storm a calm, so that the waves thereof are still.

Then are they glad because they be quiet; so he bringeth them unto their desired haven".

POWER ON TAP

BADGERS are now protected by a law which classifies them as an endangered species. There are, however, still a large number in certain areas, where they can do considerable damage by raiding gardens. During one particularly dry summer many gardeners in our neighbourhood found their carrots had been dug up and eaten by these nocturnal creatures, who left the owners nothing but the tops, all laid out in a neat row.

We were able to keep them away from our own crop in the field with the help of a little screw called a 'self-tap'. How did the screw help? Well, we had an electric lamp which not only shed a bright light, but flashed a red warning light at the same time. The tractor has a powerful battery, and by using this little screw to make a good contact between the battery and the lamp, the light flashed all night and kept the badgers away until the crisis had passed.

God has all power to enable us to live our lives aright, and good contact with him will ensure that his power will operate in us. Just as the terminal is the only place on the battery where we can pick up the electric power it contains, so the only point of contact for us with God is through the Lord Jesus. If we are like this little screw, and keep in close contact with him, we too shall be able to shine in this dark world as he has bidden us. Paul tells us, "He is able to do exceeding abundantly above all we can ask or think, according to the power that worketh in us." Let us endeavour to keep the 'self-tap' screw of our faith firmly in contact with our Lord, and thus be energised by him.

THE BADGER

The badger of the Bible
  is little read about,
Its qualities are few, but there
  is one worth pointing out:
And that concerns its useful coat,
  with fur both strong and soft.
A cover for the 'tent of God'
  which Moses reared aloft.

The badger's skin well covered up
  the things for priests to use,
And was the covering that God
  himself did wisely choose.
The ark and lamps protected were
  from moisture, heat and dust,
That all beneath its covering
  could safe abide from rust.

Another covering God chose
  to clothe the sinful soul;
Not skins all joined together,
  but a great and glorious whole.
'Tis Jesus' robe of righteousness,
  wrought out by grace and love,
To cover us and fit our souls
  to dwell with him above

LAZY STREAKS and WILLING HORSES

TEAMS of working horses are no longer common in our country; with their passing has gone a way of life and a small world of skills. When I took an old uncle to see the Sussex Shire Horses, he was reminiscent about when he was a lad and saw these magnificent horses pulling great trunks of oak on timber tugs out of the local woods. Four of these horses, pulling together, and using the tremendous 'purchase power' of their sixteen enormous feet, could move a load which would leave an eighty horsepower tractor helpless, with its wheels spinning.

Sometimes he said two teams would be put together to pull up a steep incline. It then needed special care, and a sharp eye on the part of the carter, to see that every horse did its share. Without this it would have been easy for a horse with a 'lazy streak' to walk along, just keeping pace with the others, but not 'putting his heart into it'. On the other hand there is an old saying, "You can work a willing horse to death", which sprang from the fact that this sometimes happened when the leading horse had been allowed to do more than his share.

These two extremes are still present wherever there is work to be done by a team of people. Just how those old carters managed to keep that necessary balance I do not know, it is one of the lost skills. One thing I do know is that where there is work to be done in a community or church, those who have the task of co-ordinating the workers would like to be able to motivate some of the passengers among them, and take the strain off the 'willing horses' who shoulder most of the burden. All are willing, but there are some who are willing to work, and others who are willing to let them.

The Bible says, "Whatever your hand finds to do, do it with all your might...Do it heartily, as unto the Lord."

Lazy Streaks and Willing Horses. page 53.
A pair of Sussex Shire horses in their Sunday Best.

JULY

A CHILD'S DOLL IN THE HAYFIELD

HAYMAKING has many nostalgic memories for my generation. Grass cutting was often done early in the morning while the dew helped it 'Stand up to the knife'. The sound of the rhythmic tramp of the horse's feet, the chatter of the mower, and the song of the skylark, made amends for the early start. Days later, after it had been turned by a swatheturner and rowed up with a horse rake, it was gathered on a pitch-fork, loaded onto wagons, and built into a stack ready for cutting out in the winter.

Times and methods change, nowhere more than on the farm. The day came when we were doing our first 'baling' instead of pitching loose hay; the baler was gobbling it up and punching it into stringtied bales, ready to be stored in the barn or a stack. My wife brought our two little children to see this wonder, and Rosemary brought her favourite dolly to share in the excitement. When I came home later my wife had a request, "Would you please go back to the hayfield and look for Rosemary's dolly?"

"Where did she leave it?" I asked. "She made a little nest for it in the row of hay left on the outside. It was just inside the gate on the left".

"But," I protested, "we only left that row to bale last as it was not quite ready. It's all baled up, and you know how clean that machine sweeps. We'll not see that again until it comes out all smashed up in the cow's manger."

I was unwilling to go on such a fruitless errand, and only reluctantly complied with the plea, "I wish you would go and look. I told Rosemary if she prayed about it Jesus would look after it for her." Feeling inwardly 'what a stupid thing to do' I returned to the field, entered the gate and looked to the left. Immediately I felt a strange feeling come over me. There, sitting

In the hayfield. page 55.
Baling at Cherry Gardens.

Landgirls help bring home the hay at Tilly Farm in the Sussex heartland.

unharmed in what had been the centre of the baler's path, was Rosemary's dolly. I then understood why it was said of the people who saw the miracles of Jesus, 'fear came upon them'. There is something awesome about seeing God at work; it humbles, rejoices, but never elates.

A further lesson awaited me when I took the dolly up to my little daughter in bed. She simply put out her hand and pulled it down under the bedclothes with an action which said, 'Well, I asked Jesus to take care of it, and of course he did'. I was a preacher, but my child's attitude preached to me effectively on 'With God nothing is impossible'.Do you really believe that?

IMPOSSIBLE means something inconsistent with natural laws or conditions. "It is easier for a camel pass go through the eye of a needle, than for a rich man to enter into the kingdom of God".

The camels of the wilderness
 have many useful features;
Yet all who see them will confess
 they are ungainly creatures.
Their legs are long, their feet are great,
 their backs are high and humpy,
They walk with an unusual gait,
 and are, by nature, grumpy.
A narrow passage one might try
 to pass, if thin and poorly,
But to get through a needle's eye
 is past all reason, surely.
Yet this, though but a foolish dream,
 could be a truthful story
Before one rich in self esteem
 could enter into glory.
Salvation is of God alone,
 though freely to us given,
The Saviour must for sin atone
 ere we can enter heaven.

The Royal Oak. page 59.
Picture of an evergreen oak at Danemore Farm, taken on 1st of January 1991.

THE ROYAL OAK

IN the county of Sussex, where oak trees are called "The Sussex Weed", there are a number of Public Houses named 'The Royal Oak'. Tradition has it that there was once a tree called by that name because a king, pursued by his enemies into a wood, climbed into its branches and hid among the leaves. Although the soldiers sought for him, he was safely hidden, and thus a king found shelter in a tree.

There is another tree that has proved a shelter, not only for a king, but to countless thousands beside, one that was cut down and formed into a rough cross by Roman soldiers. It is the tree to which they nailed the Lord Jesus, the Son of God, on the charge of being the King of the Jews, and seeking to set up a kingdom in opposition to the Roman Emperor. Peter tells us of that tree in his first epistle, chapter 2 verse 24, "Who his own self bare our sins in his own body on the tree".

O, Cross of Christ! O glorious tree!
What place can be compared to thee?
There God's own Son was crucified,
And for our sins a ransom died.

That tree, the cross of Christ, is the only shelter from the judgement of God against our sins. It is a tree of life to all who trust themselves to its protection, and like the apple tree Solomon found in the wood, we may sit down under its shadow with great delight., and find the fruits of salvation sweet to our taste. Song of Solomon 2 v3.

Although the Lord Jesus did not set up an earthly kingdom, he did set up a kingdom, which will never pass away. Today there are thousands of people all over the world who belong to his spiritual kingdom in fulfilment of the prophecy made to Abram, "In him shall all nations of the earth be blessed."

THINKING AND KNOWING

I ONCE bought a new piece of haymaking equipment, and as it was the first machine of its kind in the district, a very old man, who owned the farm next door, came along to see it operating. Having set the machine according to the maker's instructions, we watched it go round the field for the first time, and then I stopped the driver and made an adjustment. As it moved away I said to my friend, "I think that makes a better job of it: what do you think?"

"Think!" he replied, almost indignantly, "Think! I don't think, I know." Then he tapped the ground between his feet with a heavy stick which he always carried, and continued, "I be as sure of that as I be sure I'm standing on this 'ere bit of ground."

What a blessing it is to know! I have often thought of that old man as I read in the Bible of the testimony of those who could say, "I know" about many great truths which are in danger of being lost sight of today. Job could say, "I know that my Redeemer liveth, and that he shall stand at the latter day upon the earth." Paul, faced with difficulties and dangers which would have appalled many, said confidently, "I know whom I have believed, and am persuaded that he is able to keep that which I have committed to him against that day." My old friend's assurance was based upon years of experience which I, as a younger man, did not possess, but all, young and old, may be absolutely certain of the great truths of the Bible. They are the testimony of God himself.

It is a fact of life that we only really know what we have experienced. Many have turned to farming with a good deal of knowledge about the work, but have come to grief because of the lack of personal experience. It is the same with the things of God - we need his Spirit's witness within us.

AUGUST

THE WATER OF LIFE

LIVING in a fairly rural area it was not until the early nine-teen fifties that we enjoyed the advantages of mains water. Until that time we drew water from a well that must have served many generations, and yet had never been known to run dry. The water was brought up by a bucket attached to a very old chain, which was fixed on a wooden roller, and operated by a simple handle.

Directly after the close of the 1939-1945 war the 'Farmers Weekly' ran a 'Dustbins for Dandelions' scheme, which meant that children from East London could exchange the streets of London for the fields of the country during their school summer holidays. I well remember the reaction of one little girl who eagerly came with me to get the water. As I drew back the well cover she gave one look into the well and said emphatically, "I 'aint drinking water out of that dirty 'ole." Despite the uninviting appearance, the well tapped an underground spring and contained what the Bible calls 'living water'. Once our little friends found enough faith to sample it for themselves, they became enthusiastic about its superior qualities.

How much this is like the Living Water our Saviour has to give, as promised to the woman at the well of Sychar. To the natural mind the gospel is as unattractive as that 'dirty 'ole', but once by faith we draw water from the wells of salvation we find them superior to all other. I think also of the thousands of times that old bucket had been used, and yet each time it came up it was full of fresh water. How many believers have found that old familiar texts have proved to be like that old bucket - a means of drawing fresh water of life from the well that will never run dry.

Green Fingers. page 40
The seventh generation with his garden produce.
See The Never-ending Miracle, page17.

PLENTIFUL RAIN

A FRIEND once said to me, "It hasn't been a very nice summer has it?" I think my reply took her by surprise, for I said, "That depends whether you ask a holiday-maker or a lettuce." When we think of the terrible effects of drought, wherever it occurs, we ought to be thankful that in our own beloved land the fields are nearly always green. With the wide variety of crops grown in this country we can be sure that the weather is always just right for something.

When I was actively engaged in farming and market gardening, I always tried to organize myself so that the ground was ready to plant the various kinds of brassica about the same time as we made hay, so whatever the weather, I could keep smiling. One day an old retired farmworker said to me, "I 'aint never met a farmer like you before. I've never once heard you complain about the weather." I replied by saying that it was not the under-gardener's place to complain about what the head-gardener decided to do, and that it was my job to make best use of what the Head-Gardener sent.

The effect of a plentiful rain can sometimes be quite dramatic, even in our own country, and so much more in the near desert areas of the world. This term appears in Psalm 68.v9, "Thou, O God, didst send a plentiful rain, whereby thou didst confirm thine inheritance when it was weary." As the psalm refers to the journey of the Children of Israel through the wilderness, it would appear the plentiful rain could have been the quails which fell around their camp when they cried for meat. Later, when they were in the promised land, this was literally true, as God sent them the rain in its season.

Believers are on a journey from Egypt to Canaan in a spiritual sense, and, just like travellers in a

Well Shod by God. page 65.
Going to school along Broomfield Lane, which is mentioned on page 7.

wilderness, they have times of weariness. This may be occasioned by physical effort, sickness of body, sorrows and losses in the providential path, spiritual weariness through a sense of sin and failure, or a feeling of being far off from God. Often both the natural and spiritual element is present. How welcome then are those times of refreshing and confirmation sent from above.

As there is a plentiful rain for the earth, so there are also showers of blessing for God's children, but we must not forget that rain comes from the clouds.

"Ye fearful saints, fresh courage take,
The clouds ye so much dread
Are big with mercy, and shall break
In blessings on your head."

Are you one of the weary ones? Apply to him who is mentioned in v18 of psalm 68. Seek the showers of confirmation, and then go forward in assurance of the plentiful rain, for he has said, "My grace is sufficient for thee, and my strength is made perfect in weakness."

WELL SHOD - BY GOD

When my son's little boy started school, the fourth generation of our family to make that two mile journey, I contrasted the ease of his journey by car with the lot of those who had gone before. This brought back very vividly the time when my own first-born began school.

As my wife prepared the necessary clothes etc. for this, she said to me one day, "I can't bear to think of that little mite having to trudge to school, perhaps through ice and snow, and sitting all day with cold feet. Could we buy her some lamb's wool lined boots for the winter?" When I asked the price, and was told £5,(we were getting £3 a week wages) I said, "No, I'm sorry, but you'll have to forget it." Forget it I did, but my wife did not.

A few days before the winter term began I came home from the early morning milking to be met by my wife, her eyes bright with that light the believer recognises is of the Spirit, saying, "God has said, 'Before they call I will answer.' You remember I asked about fleecy lined boots for Rosemary: I hardly knew whether to pray for them or not because it did seem something of a luxury, but see what the Lord has sent."

In the post that morning had come a beautiful pair of lamb's wool lined boots, and although Rosemary's feet were very small, needing a narrow fitting, they were just the right size. Not to this day do we know where, or from whom they came, but their arrival confirmed the Saviour's words, "For your heavenly Father knoweth what things ye have need of before ye ask him." The arrival of the boots was a surprise, but we did not wonder that they fitted her. He who knew the number of the hairs on the head of our little one, was not likely to make a mistake in the simple matter of the size of her feet. If only we could always grasp this truth!

"If God spared not his own Son, but delivered him up for us all, will he not also with him freely give us all things?"

THE BANK MANAGER

MANY of the changes in agriculture over the past forty years are apparent to even the most casual observer. Methods and machinery, crops and cattle, have changed dramatically. Other changes are less obvious, but in some cases even greater, as in the matter of finance.

There was a time when it was the farmer's boast that 'his Bank was in his back pocket', and that he bought nothing but what he could pay for with cash. Today, with greatly increased capital expenditure and investment in machinery etc. there can be few who farm without an overdraft. Some of the largest farms carry overdrafts which would horrify our forebears.

I was brought up with a healthy horror of an overdraft or loan from the bank, and a sense of almost panic seized me one day when opening my post I found a letter from the Bank Manager saying, "I thought you would like to know that your account is overdrawn." Why he thought I would like to know that, I could not conceive! The overdraft was £280, and my immediate reaction was "What shall I do? What will my Dad say?" These questions were quickly answered, for in a few hours a local corn merchant called. He wanted ten tons of barley in one hundred-weight bags, for which he was prepared to pay cash. The price? £28 per ton!

I quickly gathered up what spare cash we had, and together with his cheque, made my way to the Bank to put the account back into the black. This may seem to be of little consequence to many today, but to me it was a matter of great thanksgiving to God, for neither before nor since did I sell corn on the spot: usually it was a case of waiting for somebody to buy my bagged corn, and then waiting at least a month for the money. I was convinced that my heavenly Father had once again shown his interest and concern for my comfort of mind. As he has said he hears the young ravens when they cry, can we not trust him to attend to our needs?

"I know who holds the future,
And he'll guide me with his hand,
With God things don't just happen,
Everything by him is planned;
So as I face tomorrow
With its problems large and small,
I'll trust the God of miracles,
Give to him my all."

THE RAVEN

"He giveth to the beast his food, and to the young ravens which cry." Psalm 147 v9.
"I have commanded the ravens to feed thee." 1 Kings 17 v4.

Everybody loves the robin
With its red and cheery breast;
And the birds that sing so sweetly,
Or that build a pretty nest.

But who cares about the ravens
With their black and ugly wing:
With their nasty unclean habits,
And who croak instead of sing?

Yet we know God cares about them,
For he hears the young one's cry;
And the Bible tells us simply
That he does their needs supply.

What is more, God used the ravens
When his servant wanted meat,
So that they supplied him daily
With sufficient food to eat.

Oh, how wondrous is our Father
Who controls all nature so;
That, in any sort of trouble
We to him in faith may go.

All the thousands of the cattle,
All the treasures of the gold,
Are the Lord's to make provision
For his children, young and old.

SEPTEMBER

LOSING THE LOAD

AT Harvest Thanksgiving we remember that God has fulfilled his promise and enabled farmers and gardeners to 'bring forth food out of the earth'. Although we no longer see the old scene of the loaded wagons bringing home the sheaves of corn, many more people are growing their own vegetables, and so carry home their harvest, sometimes on their bicycles, in their cars, or even on their backs.

One day a newly converted Christian was bringing home a large sack of potatoes, when a non-Christian friend stopped him and said, "I hear you have become a Christian. How do you know Jesus has saved you?"

The believer took a few steps, and then let the bag full of potatoes slip from his back. "How do I know I have dropped this bag of potatoes?" he asked, "I haven't even looked round."

"You haven't looked round", replied his friend, "But you no longer feel their weight on your back."

"That is one reason why I know I am saved", explained the first, "I don't carry the heavy burden of my sins any more. The Lord has taken them all away, and given me peace and joy instead."

God's word declares, 'All have sinned and come short of the glory of God'. That means me, and you, and every one of us; we have come short of the standard required for knowing God, enjoying his blessings here below, and heaven above, but 'If we confess our sins, he is faithful and just to forgive us our sins, and to cleanse us from all unrighteousness'. This is as true as the first statement, and those who come to know the first - the reality of the burden of sin - and then the joy and peace which comes by faith through believing on the Lord Jesus as Saviour, have very good reason to say, "I know I am saved."

Cutting oats with a horse-drawn binder.

The Useful Prop. page 71.
Thatching stacks where the useful prop was used.

THE USEFUL PROP

BUILDING a cornstack was a work of art, a feat of engineering! Beginning with a small round 'steddle', often made of faggots overlaid with straw, the slippery sheaves of corn had to be built into a stack which grew in circumference with each layer up to the height of the eaves. This was to prevent the rain running down the sides, penetrating into the stack and spoiling the ears of corn. Sometimes a bulge would appear as the stack progressed, and then the stacker had to resort to a prop in order to prevent a collapse.

One such prop is indelibly etched in my memory. In the autumn of 1948 I wandered one Sunday afternoon into a neighbour's corn field and leaned meditatively upon this prop. That day, for the first time in my life, the Lord had shed his love abroad in my heart by the Holy Ghost. Sovereignly, unexpectedly, and at that time unsought. Without a word being spoken, or for any apparent reason, I was immersed in the love of Christ, feeling my whole being taken up by him. I leaned upon that prop savouring the wonder and experience captured for us by Wesley in his hymn:-

"Love Divine, all love excelling,
Joy of heaven to earth come down;
Fix in us thy humble dwelling,
All thy faithful mercies crown;
Jesus, thou art all compassion,
Pure, unbounded love thou art,
Visit us with thy salvation,
Enter each poor trembling heart."

My little Welsh collie was looking up into my tear-bathed face, and I remember saying, "Flossie, there was a time when I envied you for having no soul to be lost, but now I know a joy and peace that can never be yours. "Alas, I have not always lived in that blessed frame of mind, but the memory of it has been a prop when I might otherwise have collapsed.

Do you ever thank God for the props he has given you? Memories of past blessings, friends, books, the means of grace, daily tokens of his favour. What sort of 'stacks' would we be without our props?

THE ESSENTIAL PROP

IT had been a bumper harvest, and now, in the winter, we were busy threshing the corn out of the stacks. The granary was full: where should we put the wheat? The loft in the old barn appeared the obvious choice, and soon we were carrying the $2\frac{1}{4}$ cwt. sacks up the ladder and stacking them in neat rows across the floor.

Suddenly a startled cry from my father caused me to pause. "David, David, look at that beam!" The single beam which carried the floor joists was sagging dangerously under the unusual load. Quickly, but carefully, we moved the sacks away from the middle of the floor, then getting a stout oak post, erected it under the centre of the beam as a 'prop'. The old beam was still weak and unreliable in itself, but now it was strengthened as it rested on that one prop, which took all the strain.

In Isaiah we read,"Thou wilt keep him in perfect peace whose mind is stayed on thee". I understand that the original word 'stayed' carries the implication, 'has but one prop'. What an illustration! what a promise! However weak we are, however near to breaking point we may have come, to stay the mind upon God, as that old beam was stayed on its one prop, is to prove for ourselves his promise to Paul, "My strength is made perfect in weakness". Though in ourselves no stronger, we shall not only bear up under the strain, but enjoy that perfect peace,

"Thou wilt keep him in perfect peace...because he trusteth in thee. For in the Lord Jehovah is everlasting strength".

RISEN FROM THE DEAD

ONE of the diminishing features of country life in Sussex is the pheasant shoot. Fifty years ago there were several estates in the locality where pheasants were reared, and during the autumn and winter men and boys engaged to drive the birds over the guns. The saying was, Up goes a sovereign, (cost of rearing the bird) Bang goes sixpence, (the cartridge) Down comes half a crown, (price of the bird in the shop).

Among the men who regularly went 'beating' was an old character known locally as 'Codger'. He was a very old man, so we were not surprised to hear that he had died. However, when the shooting season began again, I was walking through the woods to the starting point, and who should I see coming through the trees, making for the path I was on, but old Codger. My first reaction was one of fear and disbelief, but I soon discovered it was the old man in living flesh.

On returning home at the end of the day I said to my father, "Old Codger isn't dead, he was at the shoot today." He replied, "You must be mistaken: it was probably somebody who looked like him." Nothing I said could convince my father that old Codger was alive, and certainly nothing would ever convince me that I had been mistaken. Not only had I seen him and spoken with him, but we sat down together with the others eating our midday break of cheese sandwiches.

This experience has enabled me to understand the feelings, first of doubt and fear, and then the absolute conviction displayed by the disciples at the resurrection of the Lord Jesus. Peter said, "Him God raised up on the third day, and showed him ...to us, who did eat and drink with him after he rose from the dead."

Risen From The Dead. page 73.
A typical "old Codger" rises from his meal in the woods where we were beating.

OCTOBER

WIND IN THE OAKS

THE hurricane which struck the southeast corner of Britain on October 17th 1987 will remain in the memory of all those who witnessed it. Thousands of mature oak trees were uprooted, for being full of leaf they presented a large target to the wind, and as the ground was excessively wet, their hold in the earth was greatly weakened. Nevertheless it was an awesome scene of destruction. The thing which most impressed me was to see a lone oak over a metre and a half thick, and in a field where its roots still held firm, snapped halfway up its massive trunk as though it had been a matchstick.

At the height of the storm I went out and rescued my car from a building which was on the point of collapsing, and so felt personally the full force of the wind. Reflecting upon this afterwards I thought of the Lord Jesus standing up in the boat on the lake of Galilee and stilling such a wind.

The Bible says of our God "He bringeth the wind out of his treasures", and in the Book of Proverbs we read, "He gathereth the wind in his fist". This gave me a wonderful mental or metaphorical picture of God opening his hand and letting the wind blow, and the Son of God opening his hand, and gathering it in his fist. If only we could always live in the knowledge of the almighty power of our God, 'Who maketh the clouds his chariot, and walketh upon the wings of the wind'.

THE WICK LAMP

IN my living room I have two oil lamps. The Aladdin lamp we used before electricity came to us about 1953, and an ordinary wick lamp with a glass bowl recently given as 'being in keeping with the house'. The latter

Wind in the Oaks. page 75.
Buildings which were devastated in the storm.

reminds me of the lamp used in the same room when I was a boy, a double burner, that is a lamp with two wicks, which was the ultimate in lighting for country cottages during Victorian times. This lamp had a glass bowl for the paraffin oil, so it was easy to see when the fuel was in need of replenishing. It also had a glass 'globe' which was decorated with a white lace pattern, and fitted round the funnel, or chimney, so hiding the flame and dispersing the light.

There was an art in using the oil lamp. The funnel which guarded the flame from draughts must be kept clean. The oil itself must be genuine, of the right kind. To use a very light oil, such as petrol, would cause an explosion, while heavy oil would not only produce a feeble flame, but give off fumes and smoke. It would also 'char' the wick, that is actually burn the wick, turning the end into a black charcoal-like mess.

The same charring would occur if the oil became low, and the light would go out completely when the container ran dry. The normal way of putting out the lamp was by using the shutter, a device operated by a lever on the side of the carriage, which made a little brass plate suddenly clamp over the top of the wick and shut off the flame. The carriage was the housing for the wick, which could be turned up or down by a wheel. The flame itself was interesting to observe. Those who are familiar with a gas cooker will have noticed that the flame actually dances a little distance above the holes through which the gas comes. The same principle is at work in the oil lamp. When a flaming match is applied to the end of the wick, the oil, drawn up the wick by capillary action, evaporates into gas, and burns above the wick. So long as there a good supply of oil, the wick itself never burns.

"Give me oil in my lamp, keep me burning,
Give me oil in my lamp I pray;
Give me oil in my lamp, keep me burning,
Keep me burning till the break of day."

ALADDIN'S LAMP

THE Aladdin lamp carries the wick lamp principle a stage further. Instead of two straight wicks, two wicks are woven together for half their length making a cylinder which fits inside a round carriage, and is moved up and down by turning a winder mechanism similer to that used in the old type. When this round wick is lit, then the funnel must be placed over it. In the base of the funnel is a 'mantle', a conical device made of white lime-like substance, very delicate and intricate, and suspended on a wire frame. Having put this securely in place, the operator turns up the wick, and, wonderful to behold, the flame itself seems to disappear, leaving the mantle glowing brightly - a miniature 'limelight'. I have a great affection for this form of lighting, and at one time used it in my study instead of an electric reading lamp.

When Jesus said in Matthew 5.16,"let your light so shine before men that they may see your good works and glorify your Father in heaven", he had a lamp in mind. The R.A.V. rightly renders v15 as 'light a lamp' and 'on a lampstand'. The lamps of his day were not so sophisticated as those I have been describing, but the principle on which they worked was exactly the same.

There are many lessons to learn from the lamp. The oil reminds us of the Holy Spirit, who is represented by this figure in the Bible. The wick of our inner self must be free from any sort of encrustment which hampers the free flow of the Spirit. The glass of our daily life must be clean, so that the light of our witness is not marred. Cleanliness in every part of the lamp is essential or there will be unpleasant fumes, which should also remind us that only as our lives are sanctified through the truth can we be 'a sweet savour unto God'. The flame must be imparted from outside the lamp itself, just as we must receive the life of Christ if we are to shine as Christians.

There is a sad lesson too - how often do we, by our sin, 'shut off 'the supply of the Spirit?

As the humble wick lamp gives us lessons how we should shine in this present world, so the more refined Aladdin lamp, with its pure white mantle, should be a reminder that in the world to come the Christian will shine in a pure white mantle, the robe of Christ's righteousness.

THE OIL OF THE SPIRIT

YOU'll crack up man". These words were going through my mind as I came home from Tunbridge Wells one day in the midst of a strenuous time. They had been spoken to me by a dear brother whose love and concern prompted him to say, "If you go on like this, you'll crack up man". There were days when I could not trust myself to drive those six miles without pulling off to the side of the road and putting my head down for a while to overcome the tiredness which beset me, but on this occasion I was doggedly driving on. Not far from Cherry Gardens there stands a railway bridge about a hundred yards from the river bridge. As I drove the short distance between those two bridges the Lord gave me an experience I doubt if I shall ever forget.

"You'll crack up man". Was it my friend's voice, or the voice of the Tempter? Suddenly in my mind's eye I saw the old glass bowled lamp. I saw it empty of oil, the wick charred, burnt out, and I saw the bowl full, the flame clear of the wick, and the light shining brightly. A tremendous thrill ran through my whole being. My physical tiredness vanished, my jaded spirit was replaced with a humble confidence in Him who enabled Paul to say, "I can do all things through Christ which strengtheneth me." I said, "Thank You Lord; Keep me filled, and I shall be able to keep going." Is there a believer who is 'burning out'

Beauty in Decay.

a believer who is 'burning out' instead of 'burning bright'? Some poor soul tempted to give up, jaded in spirit, taunted by the Devil, and feeling useless and despondent. May the Lord bless you through the message of the lamp. Turn to Him of whom John said, "He shall immerse you in the Holy Ghost and with fire". Do not fear to ask him to cleanse your whole being, empty you of every other oil, and fill you with His Spirit. In the warmth generated by his love the flame will burn more brightly, as the wick of your poor soul becomes the instrument through which the oil of His Spirit and grace flows.

May your prayer and experience be:-

Jesus fill now with thy Spirit
Hearts that full surrender know,
Cause the streams of living water
From our inner man to flow;
Channels only, blessed Master,
But with all thy grace and power
Flowing through us, Thou canst use us,
Every day and every hour.

BEAUTY IN DECAY

I saw a holly leaf one day,
Its ribs were stark and bare;
The flesh had fallen quite away -
Yet I saw beauty there.

Whatever be our outward dress,
Or beauty of the face,
May God's good Spirit give to us
The inward light of grace.

That when our outward form declines
Those near to us may say,
The inner light more brightly shines -
There's beauty in decay.

SEASONS.

Behold the trees in winter stand
As if they're touched by death's cold hand,
And in that season of the year,
No life at all does there appear:
Be pleased , O Lord, thy power to show,
May we thy quickening influence know.

Then later on a change is seen,
The trees all stand bedecked with green,
And brave new growths all round appear,
Whilst flowers declare that spring is here;
May we beneath God's showers of grace
Show forth his praise in every place.

In summer when the fruit we see,
'Tis then we note the useful tree;
We find, if we the tree should tap,
The root is full of living sap;
May we possess a lively root,
That we might bear some goodly fruit.

In autumn glorious tints are seen,
Gold, amber, red, where all was green,
And yet, for all this varied show
The living sap has ceased to flow;
May we, though leaves should fall away,
Be saved, O God, from root decay.

Though we through various seasons pass,
As changing as the weather glass,
Yet we would bless thy holy name
That thou, O God, art still the same.

NOVEMBER

REFLECTIONS

IT had been a lovely day for November, warm and sunny in the middle of a wet spell. I came in from work as night fell and got ready to go out for a preaching engagement. When the time came to start the rain was falling, giving the prospect of an unpleasant journey through leaf-strewn lanes. As I was getting petrol at a local filling station I could feel the cold rain on my head, penetrating where the 'thatch' had grown thin. "Not a very pleasant end to a lovely day", I said to the attendant as I went to pay. He looked up quickly and said, "The day isn't ended yet sir."

The day isn't ended yet sir! Journeying on my way I reflected on those words. What a wealth of meaning there was in them, especially for the Christian. Not only was it literally true (there could have been a beautiful star-lit sky before midnight) but it is true of the stormy periods of our lives. I hope that if ever I am tempted to murmur or begin to despond because the clouds gather over my pathway, I shall hear the quiet voice of that pump attendant, "The day hasn't ended yet sir."

Indeed for the Christian, however dark or bright the path may be, the best is yet to come, a day unclouded by sin and sorrow, but bright with the beauty of holiness. Are you, my fellow believer, tempted to give up? To say, as you come toward the evening of life, "I have had a good day, but my present circumstances are bringing it to a not very pleasant close"? Does Satan tempt you to despond? Then look up, and say, "The day isn't ended yet sir." Your heavenly Father has given a precious promise for your encouragement, "At even time it shall be light."

Beggar's Bed. page 85.
This hollow oak on the outskirts of Crowborough was once home to a local tramp.

A BEGGAR'S BED

SOMETIMES on a winter's night, when the gale howled outside and we were sitting around the fire, busy with various activities, my father would say, "Here we are, warm, safe, and sound, while out there is some poor tramp holed up under a hedge."

The local tramp was quite a feature of country life at one time, some apparently well educated, others dim-witted (but sharp enough when looking out for Number One!). According to the law they had one thing in common, they 'had no visible means of support, and no fixed place of abode'. During the winter they would take refuge in all sorts of strange places and buildings, even the 'little house' at the bottom of the garden which was the countryman's toilet. When they began to make a nuisance of themselves they would often be hauled up before the magistrate on the above charge, and sent to jail until the spring came.

It is sad to reflect that they were not only reduced financ- ially, but that they were content to live that sort of life. Sometimes on a fine but cold winter's night they would make use of the heat generated by a farmyard manure heap, by wrapping themselves in a greatcoat, and lying down without apparently noticing the smell, or the degrading nature of their bed.

It is said of God, "He raiseth up the poor out of the dust, and lifteth the beggar from the dunghill, to set them among the princes, and to make them inherit the throne of glory." (I Sam. 2.8) Perhaps our generosity would go as far as giving such a character some food, but would we be prepared to invite such into our houses? I wonder how many would do as a man I knew, and put a notice on the door, "To Tramps. Come in brother, and help yourself"?

We may pride ourselves that we are not so degraded as to lie on a dungheap, but how many there are that lie on a bed of sin, which is far more degrading in the

Continued
on page 89

A White Christmas? page 87.
Snow on Christmas cards is pretty, but its pretty awful at times for those living out in it!

A WHITE CHRISTMAS

"Will there be a white Christmas?" How often have I been asked that question, and what different reasons have prompted it. To some snow comes as a welcome interlude, for what child or young person does not in some degree like snow! Making a snowman, sliding or sledging, are great fun for those who are young, and even among the more elderly there are those who say, "I always feel so well when there is snow on the ground."

To those who still earn their daily bread by the sweat of brow in farming and gardening, the snow is, in many ways, unwelcome, and we look forward to the time when the earth will be free from the grip of winter. Yet even to the tender plants the snow may bring benefit. Often I have seen rows of broad beans or peas, planted early in November, look fine and green until bitter wind and frost have come. Where the snow lay thick, the plants would still be green and thriving, but where there was none, they would be black and frost-bitten.

So often we prove in our lives that God allows pain, suffering and difficulties to come like a blizzard, threatening destructtion. We long for all sunshine and ease, but He who saw the necessity for winter in nature, also sees the necessity for these things in the lives of his children, that by them we may not only have our characters strengthened, but prove all things work together for good to those that love him. Often the very things that give us pain and distress are a means of saving us from greater harm and sorrow.

"God has not promised skies ever blue,
Flower-strewn pathways all our life through;
But He has promised strength for the day,
Grace all sufficient, help by the way."

Safe as Houses. page 89.
-And a Name to Trust? A children's playhouse, with two of my grandsons peeping out.

sight of God; indeed, none of us are free from sin's taint. Yet such is the love and grace of God that he takes sinners such as we and makes us his children. "Heirs of God, joint heirs with Christ", inheriting the throne of glory.

AS SAFE AS HOUSES

IT was what some might call a filthy night, cold, dark, and pouring with rain. As I neared home the headlights of the car picked out the figure of a middle-aged woman, with a large suit case in each hand, struggling with bowed head against the storm force wind. It was evident she had just got off the train, and having missed the last bus, was having to walk home.

In a few minutes I had stopped the car, opened the rear door, and said, "Put your cases in there, and I will take you home." Gratefully she responded, and we were on our way, but her whole attitude changed when she realised she was in a car with a total stanger. "Where do you live?" I asked, as we almost immediately came to cross roads. Hesitatingly she told me, and then asked where I lived. On being told this her tension visibly increased, because we were now going in the opposite direction to my home. Then she said, "I used to work in the Post Office, perhaps I would know your name?"

"I'm David Obbard of Cherry Gardens", I said, and immediately she relaxed back into the seat with a sigh of relief. "Thank God", she said, "then you're as safe as houses." Although I was personally unknown to her, the good name of my forebears was sufficient to allay her fears. This gave me an opportunity to mention another, and far greater name, which has brought peace, blessing, and freedom from fear to countless thousands:-

"How sweet the name of Jesus sounds
In a believer's ear.
It soothes his sorrows,heals his wounds,
And drives away his fear."

She felt as safe as houses, but houses have often failed to protect those who have put their faith in them. My father's house, generations of 'Chapel People', was a worthy one, but I might not have been worthy of the same trust. Those who trust in the name of Jesus are not just 'safe as houses', but safe for eternity.

THE ONE WAY TICKET

There are but one way tickets
  as you pass along life's way,
And no more will you travel
  the path you tread today.

If you can do a kindness
  to a fellow mortal here,
Or by a word, or kindly look,
  the heavy-hearted cheer;

If you can help another
  when your heart is full of pain,
Or show a patient spirit
  in the midst of stress and strain;

If you can counter anger
  with your gentleness and love,
Yet stand against aggression
  in the name of God above;

If you can help the weakling,
  give the poor a kindly hand,
The young, the old, the burdened,
  or stranger in the land;

Oh, do not fail to do it
  as a favour God bestowed,
For no more will you travel
  on this portion of the road.

DECEMBER

ONE STEP AT A TIME

AT the age of 67 I fulfilled a long-cherished ambition to walk the South Downs Way, from Eastbourne almost to Petersfield. On hearing of this intention a friend said, "However are you going to walk eighty miles?" My reply was simple; "I shall just try and keep putting one foot in front of the other. I only need to take one step at a time".

That conversation reminds me of a story I once heard concerning a little boy who was looking into an old barn in the fading light of a winter's afternoon. Someone was moving about in the hay loft, and the little fellow came to the foot of the ladder and called out, "Is you up there, Grandpa? I want to come up." Grandpa replied, "All right then, come on up." The little foot paused at the bottom of the ladder, and a troubled face was lifted up toward the dim loft.

"Grandpa, I can't see the top step". This was met by Grandpa's reassuring laugh. "Put your foot on the rung where you are, little man, and climb up. The last step is here; you'll see it when you get to it." Thus assured Freddy mounted the ladder and soon found himself lifted safely into the loft by Grandpa's strong arms.

How we need to learn this lesson over and over again. So often we want to see the end from the beginning, but the path of faith is one step at a time. We must take the first step shown to us, and trust God for the next. He will give us sufficient light to see, but not to be over-confident. He seldom gives us to see the end from the beginning, but he always assures us that all will work for good.

"I do not ask the distant scene to see;
Hold thou my hand, one step enough for me."

One Step At A Time. page 91.
Where the Southdowns Way crosses the river Adur, with the church of St. Boltophs in the background.

WHITER THAN SNOW WHITE

WE do not always have a white Christmas, but when the snow falls in moderation at this time of the year there are few that do not appreciate its beauty. Not only does the snow provide fun and games for those who are still young enough to enjoy it (and fun and games of another sort for motorists), but it has its beneficial effects. It brings down a certain amount of fertilizer collected from the atmosphere.

Those of us who live in the country see the beauty that the snow brings, transforming trees and hedgerows as it freezes on the twigs. Even an ugly site or a piece of waste land becomes a thing of beauty as the pure white snow settles smoothly, or is blown by the wind into weird and wonderful shapes.

In Psalm 51 David prayed,"Wash me, and I shall be whiter than snow". He knew that God, who sends snow to benefit the earth and transform it, could also transform him, cleansing his heart and benefitting his life. He knew also that the God who causes each snowflake to differ from another, deals individually with his children, for although we have the same basic nature, no two of us are alike. Our times and paths are very different from those of David, but like him, we were all born with a sinful nature.

Have you had the transforming experience David speaks of in this psalm? Will you stand in the presence of God whiter than snow - washed in the blood of the Lamb?

WASHED IN BLOOD

THE first time I saw a cow with a prolapsed womb I had a great shock. There in the gutter was a great mass of delicate tissue, and I immediately sent for the Vet. I was in for a second shock when, after throwing a

Whiter Than Snow. page 93.
Milk churns outside the dairy door have their lids transfomed into top hats.

bucket of soapy water over the womb, he began to put it back inside with some bits of straw and dirt on it. When I remonstrated about this he just said. "Oh, the old cow will look after that. Whatever we do it will be covered with germs and bacteria." He was, in effect, saying that the blood which circulated through the cow's body was the very best cleansing agent of all.

When the Bible speaks of being washed in the blood of the Lamb, it means, as one modern version puts it, that through the merit of his sacrificial death, our sins have been dealt with, so we are no longer defiled by them.

TRUTH STRANGER THAN FICTION

THERE are many legends about the birth of the Lord Jesus which we know are not true, some of them quite fanciful. It used to be said that if anyone went into a stable at midnight on Christmas Eve, they would find the animals on their knees as if worshipping the baby Jesus. In many years of working with animals I have often been attending them at that time, 'racking up' as we called it, and can assure you they do not adopt any different attitude for Christmas Eve. The fact is that many of these creatures sleep on their knees!

There is something very satisfying about caring for animals at any time, but there have been a number of occasions when I have left the Christmas celebrations and gone into a cowshed or stable, and felt something indescribable as I have stayed to meditate on the wonder of the Son of God being born in a cattle shed. A greater wonder is that he may be born again in us.

"How silently, how silently, the wondrous gift is
given
So God imparts to human hearts the blessing of his
heaven.
No ear may hear his coming, but in this world of sin
Where meek souls will receive him still the dear
Christ enters in."

FINAL JUDGEMENT

FOR many generations Britain has been a great centre for cattle-breeding. One of the reasons for the confidence placed in British stock has been the very stringent measures taken to stamp out the dreaded Foot and Mouth disease. Not only is there strict control on the movement of livestock, but all animals that could possibly have been in contact with infected ones are slaughtered, and their carcases burned on the fire.

Several years ago, when it was necessary for us to have a permit to walk our animals along the road to their grazing field, the local policeman came to see if we possessed one. My father had gone to Crowborough Police Station to obtain this, but having mislaid it, was unable to convince the constable of that fact. Telling my father, "The fat will be in the fire now", he cycled up to Crowborough to make a charge against him. When the constable made his accusation, the officer in charge turned to the record book, and there was the evidence that my father was 'not guilty'.

There is a wonderful lesson to be drawn from this. We must all appear before the judgement seat of God, and even now may feel justly accused of sin; guilty of not keeping God's holy law. The message of the gospel is that at the very place of final judgement there will be found, for all who truly believe in the Lord Jesus, and trust him for their salvation, the evidence of their being free from all charge. This record is not written in ink, but in the precious blood of Christ, for the scripture affirms 'through this man is preached unto you the forgiveness of sins, that whosoever believeth in him shall be justified from all things'. Satan, the 'accuser of the brethren', will be silenced in that day, for, "Who is he that condemneth? it is Christ that died, yea rather, that is risen again, who also maketh intercession for us".